T0273123

Costs and Benefits of Greenhouse Gas Reduction

Costs and Benefits of Greenhouse Gas Reduction

Thomas C. Schelling

The AEI Press

Publisher for the American Enterprise Institute

WASHINGTON, D.C.

1998

ISBN 9780-84477-114-4

©1998 by the American Enterprise Institute for Public Policy Research, Washington, D.C. All rights reserved. No part of this publication may be used or reproduced in any manner whatsoever without permission in writing from the American Enterprise Institute except in the case of brief quotations embodied in news articles, critical articles, or reviews. The views expressed in the publications of the American Enterprise Institute are those of the authors and do not necessarily reflect the views of the staff, advisory panels, officers, or trustees of AEI.

THE AEI PRESS
Publisher for the American Enterprise Institute
1150 17th Street, N.W., Washington, D.C. 20036

Contents

Foreword

This volume is one in a series commissioned by the American Enterprise Institute to contribute to the debates over global environmental policy issues. Until very recently, American environmental policy was directed toward problems that were seen to be of a purely, or at least largely, domestic nature. Decisions concerning emissions standards for automobiles and power plants, for example, were set with reference to their effect on the quality of air Americans breathe.

That is no longer the case. Policy makers increasingly find that debates over environmental standards have become globalized, to borrow a word that has come into fashion in several contexts. Global warming is the most prominent of those issues: Americans now confront claims that the types of cars they choose to drive, the amount and mix of energy they consume in their homes and factories, and the organization of their basic industries all have a direct effect on the lives of citizens of other countries—and, in some formulations, may affect the future of the planet itself.

Other issues range from the management of forests, fisheries, and water resources to the preservation of species and the search for new energy sources. Not far in the background of all those new debates, however, are the oldest subjects of international politics—competition for resources and competing interests and ideas concerning economic growth, the distribution of wealth, and the terms of trade.

An important consequence of those developments is that the arenas in which environmental policy is determined are increasingly international—not just debates in the U.S. Congress, rulemaking proceedings at the Environmental Protection Agency, and implementation decisions by the states and municipalities, but opaque diplomatic "frameworks" and "protocols" hammered out in remote locales. To some, that constitutes a dangerous surrender of national sovereignty; to others, it heralds a new era of American cooperation with other nations that is propelled by the realities of an interdependent world. To policy makers themselves, it means that familiar questions of the benefits and costs of environmental rules are now enmeshed with questions of sovereignty and political legitimacy, of the possibility of large international income transfers, and of the relations of developed to developing countries.

In short, environmental issues are becoming as much a question of foreign policy as of domestic policy; indeed, the Clinton administration has made what it calls "environmental diplomacy" a centerpiece of this country's foreign policy.

AEI's project on global environmental policy includes contributions from scholars in many academic disciplines and features frequent lectures and seminars at the Institute's headquarters as well as this series of studies. We hope that the project will illuminate the many complex issues confronting those attempting to strike a balance between environmental quality and the other goals of industrialized and emerging economies.

CHRISTOPHER DEMUTH
IRWIN M. STELZER
American Enterprise Institute
for Public Policy Research

Costs and Benefits of Greenhouse Gas Reduction

Fifteen years ago, hardly anyone had heard about the potential climate changes that might be induced through greenhouse warming as a result of the burning of fossil fuels. And yet in 1992, the largest international official gathering that had ever taken place anywhere in the world happened in Rio—more than 25,000 hotel rooms were required—and a key item on the agenda was greenhouse emissions and climate change.

Some may deplore the evident lack of serious concern among the major countries of the world about addressing the problem of carbon emissions. In my view, however, five years after Rio is much too soon to be disappointed in the lack of progress. This issue will probably be with us for a century. Five years would be a very short time for any group of nations, especially 165 of them, to figure out not only what to do about the subject but even how to think about it. In addition, we must consider what kind of organization may be required and what kind of diplomacy may work.

Those afraid that too much will be done in the near term should not worry, and those who are afraid that nothing will be done in the long term should not worry, either—yet.

This essay was originally presented as a lecture at the American Enterprise Institute on November 20, 1997, based on an article titled "The Costs of Combating Global Warming," published in the November–December 1997 issue of *Foreign Affairs*. It was revised following the December 1997 conference on climate change in Kyoto, Japan. Portions of the discussion session following Professor Schelling's lecture are reprinted here.

Divisions among Nations

Two interesting divisions arose recently, one between the Europeans and the North Americans and Japanese, the other between the developed countries and the developing countries. The Europeans claimed—and it was difficult to know how sincere they were—that they would like to commit themselves to reducing carbon emissions to 15 percent below the 1990 level by 2010 or 2015. I do not think they knew how to do that. President Clinton talked about committing the United States to trying to reduce emissions to some 5 percent below the 1990 level sometime between 2008 and 2012. The Japanese were generally in line with the U.S. position.

This difference appeared irreconcilable. European governments may have been taking their extreme position knowing that they would not have to agree to any such thing internationally because the United States and Japan would not go along. They may have been comfortable advocating a position that they would not have to enforce. (Even Margaret Thatcher once proposed that she would reduce carbon emissions 5 percent below the 1990 levels by around 2005—and Margaret Thatcher knew that she would not have to stand in front of the House of Commons in the year 2005 and explain why they had not met that goal.) European leaders may have been feeling the same way— that this is a great position to take but that their successors will have to figure out whether to meet the commitments and, if so, how.

The upshot at the Kyoto conference was that developed nations—roughly the nations of the Organization for Economic Cooperation and Development—appeared to agree to reduce emissions by around 2010, on average, to some 5 percent below the levels of 1990. None gave any hint of how that might be accomplished; none appeared to have given much thought to how that might be accomplished; none had advertised to their citizens what such an accomplishment might entail. The U.S. government was laboring under a Senate resolution that precluded American participation in any carbon-emission program to which the developing nations—China, India, Brazil—were not fully dedicated. The developing nations gave no encouragement. The Clinton administration indicated it would withhold a request for ratification for at least a year, ostensibly to permit diplomacy with developing nations to bear fruit. Nobody anticipated any fruit. And the Europeans would be off the hook if the Americans stayed out because China and India stayed out. At the very best, things will be delayed a year or two, and the likelihood of reducing emissions by 2010 or 2012 to below 1990 levels—far below where they would ordinarily be in 2010!—seems remote.

The Developing World and Climate Change

The division between the developed countries and the developing is more fundamental. The developing countries' representatives typically assert that the greenhouse problem scares the developed countries and that the developed countries want to retard their development so that they will not emit fossil fuels in the way that the developed countries have done for fifty years, as their economies grow. They insist that if anything is to be done about reducing carbon emissions, the rich countries will have to pay for it. I agree with their conclusion. If anything is done, the rich countries will have to pay for it, at least for the first thirty or so years.

I disagree with the notion that the developed countries are the ones that have much to lose from climate change. The exact opposite is true. Nearly all the benefits of slowing climate change will be enjoyed by the descendants of the people currently in the developing countries.

Why? First, nine-tenths of the world's population will be in developing countries by the time climate change becomes a serious phenomenon, if it does. Second, most developed economies are virtually immune to the effect of climate change. If we ask what difference it would make to various economic activities—from automobile assembly to open-heart surgery to electronics production—it does not much matter whether we are in Washington, Oregon, Louisiana, South Carolina, or New Hampshire. The climate makes very little difference to economic productivity, especially leaving aside agriculture.

A word on agriculture: agriculture in most developed countries is a very small part of the economy. In the United States, it is roughly 3 percent. And there are so few people who earn their living mainly from farming that now the Census Bureau no longer counts them.

If the impact of a climate change on agriculture were seriously adverse and the cost of raising food went up by 20, 30, 40, or 50 percent—an unlikely event because improvements in agricultural productivity over the next fifty years ought to continue—the cost of living would have gone up 1 percent or 2 percent, while our per capita income would have doubled. Instead of reaching a doubled per capita income in the year 2060, then, we would reach it in 2061 or 2062. Nobody could tell the difference.

The developing world is different. Now developing countries are much affected by weather and climate and so are potentially vulnerable to climate change in a way that the developed countries are not. That should concern them. Their best defense against climate change, however, is their own continued development: reducing their dependence on agriculture and other weather-sensitive activities. My expectation, then, as well as my advice, is that they should not divert substantial resources from their own development into trying to slow down climate change.

Uncertainties

The uncertainties are significant. The climate change is expected to be driven partly by an average warming in the surface atmospheric temperature. "Global warming" is a misnomer because those who project climate change do not predict that temperatures will simply rise. Where temperatures do rise, they will probably rise more in winter than in summer and more at night than in the daytime. But what will primarily affect the climate will be a combination of a change in the average surface atmospheric temperature and the change in the temperature gradient between the polar regions and the tropical regions.

Some might suppose that if the climate becomes warmer toward the poles and less warm toward the equator, such changes would be desirable. But the circulatory phenomena do not affect Earth's weather uniformly: climates change in various ways. Some areas may get warmer, some may get cooler, some may get wetter, some may get drier, some may get cloudier, and some may get sunnier. To call this whole range of changes "global warming" is to oversimplify. The average rise in global surface atmospheric temperature caused by doubling the concentration of carbon dioxide—and a doubling will undoubtedly occur sometime toward the middle or the second half of the coming century—was estimated eighteen years ago at between $1\frac{1}{2}$ and $4\frac{1}{2}$ degrees Celsius.

That estimate has not been officially changed, partly because no one wants to stick his neck out with a flat prediction of a new range of error. At least officially, the uncertainty has not been reduced, a fact I find strange. With the amount of money that has gone into the atmospheric sciences, the oceanic sciences, and the glaciology of Antarctica in the past ten years, why haven't the uncertainties been reduced?

Climate may be like genetics or the human brain: something that proves to be more complicated upon further study. We did not take clouds seriously fifteen years ago, for example. Now clouds are considered a major source of attenuation or aggravation of climate change, because clouds primarily either reflect incoming sunlight or absorb outgoing infrared radiation, according to the size of the droplets and the elevation of the clouds and even according to whether the clouds are over land or ocean.

The oceans are also considered an active part of the circulatory system. Fifteen years ago, though, the oceans were seen simply as a huge cooling reservoir that would slow down the process of warming but otherwise would not have much effect.

For whatever reason, the basic estimate of the average effect of increased carbon dioxide has not officially changed. I think that unofficially the estimate has come down. The range of uncertainty may be about the same, but people refer not so much to $1^1/_2$ degrees to $4^1/_2$ degrees but to $1^1/_2$ degrees to 4 degrees—maybe, say, between 2 degrees and 3 degrees. But, at least officially, that uncertainty has remained the same.

What about the other uncertainties? A bigger uncertainty is translating a change in average global atmospheric temperature into climate change. That calculation requires understanding what would happen to the Gulf Stream and everything else. To give an example, none of the atmospheric modelers who deal with climate change has found a way to put mountains into their models. Some might think that climate in the high mountains is of little consequence. After all, not many people live at the top of big mountains. But India, Pakistan, Bangladesh, and Burma all depend for their agricultural irrigation on snow in the Himalayas the way California depends on snow in the Sierras and the Rockies.

The uncertainty of how mountains affect climate illustrates how far the climatologists are from being able to predict, region by region, or locality by locality, what the effect of increased carbon dioxide may be.

A third uncertainty, one that has impressed me ever since I got involved, has received little attention. That is, what will the world be like when the climate has changed enough to be noticeable?

If seventy-five years ago climate change had received the kind of attention it does now, what environmental problems might people have foreseen that would be either aggravated or ameliorated by climate change?

The answer I come up with is mud. Automobile tires were about three inches in diameter. When cars got stuck, horses had to pull them out. Bicycles could not be ridden in mud. People walked long distances then, and it is arduous to walk in mud. People would have said, "If climate change will make things drier, it will be a huge benefit. But if climate change makes things wetter, it will be awful."

They might not have foreseen that the country would be virtually paved solid at the end of the century.

Health and Climate

If the climate change occurred overnight—not forty, fifty, sixty, or seventy years from now—probably the most deleterious impact on Earth would be the increase in the incidence of tropical diseases, especially vector-borne diseases like those carried by mosquitoes, fleas, lice, and so forth. Malaria, dengue fever, and river blindness would all increase with the expansion of tropical areas. As it got warmer and more moist at higher latitudes, the area in which these diseases are endemic would enlarge. In addition, warmer climate tends to invigorate both the vectors and the bacteria that bring about those diseases.

If we think ahead seventy-five years, however, we have to think about how things will have changed with respect to tropical diseases. Singapore, for example, has no malaria to speak of. Across the causeway, a mile away in Malaysia, malaria flourishes, and people die. In Singapore, the rare case of malaria gets modern medical treatment. In Malaysia, many who get malaria are severely afflicted, partly because medical care is not available and partly because they are not as well nourished and healthy as the people in Singapore.

If in seventy-five years people are healthier and better fed and have better public health infrastructure, better sewage, cleaner water, and better environmental protection against the places where mosquitoes breed, we might expect that Malaysia will be where Singapore is now. That is not a very extravagant projection. It took Singapore only thirty years to get to where it is now, and if Malaysia, Indonesia, and other such countries can get there in thirty, forty, or fifty years, they will probably enjoy the kind of health that is enjoyed in Singapore, in a climate identical to Singapore's.

At a recent conference on disease and climate, I learned something quite interesting about dengue fever, which was later confirmed in an issue of *Science*.

Dengue fever is rampant along some parts of the Rio Grande. There are two towns about the same size opposite each other on the river: one on the American side, one on the Mexican side. The Mexican town, in one season, had 2,000 cases of dengue fever per 100,000 in the population; the American town had seven. With an identical climate and identical terrain, the difference in living standards, public health infrastructure, and environmental protection resulted in a factor of 300 difference in the disease. Since this is a poor American town, not a wealthy American town, I think it suggests clearly that with economic progress, susceptibility to debilitating diseases should surely go down.

To give a final example, many vector-borne diseases are severe because they afflict the undernourished, the unhealthy, and the chronically ill. A million children a year die of measles. When I was a child, I never heard of anybody's dying of measles. My children would not die now because we have vaccine. But before we had measles vaccine, only those on the margin of society ever died of measles, and we would never hear about it. The difference today is primarily nourishment. A healthy child who gets measles is quarantined for a week and goes back to school. An unhealthy, undernourished child can die.

If in fifty or seventy-five years living standards go up, with a corresponding increase in public health and medical technology, our grandchildren's grandchildren may well live in much healthier times than now.

Many of these tropical diseases are coming under fairly good control. The horrible creature that was responsible for the deaths of those people up above the Arctic Circle, described in *Smilla's Sense of Snow,* was something known as the Guinea worm. The Guinea worm is now fairly confidently expected to follow smallpox into extinction. It may well be that many infectious diseases will be much less threatening at the time in the future when we might be afraid that climate change would aggravate them.

Sharing Burdens

We should not be too disappointed if it takes a while to develop a worldwide carbon dioxide regime; nothing like that has ever been attempted, let alone successfully brought into effect. It is in no individual country's interest to spend money to do anything about climate change. It may barely be in the interest of a collection of countries as big as Western Europe to do something about it.

The burden that would have to be shared is large—not enormous, but large—especially in relation to what people can agree on diplomatically. We have no accepted standards of fairness. Shortly after World War II, countries considered a number of institutional arrangements relating to sharing burdens among countries with different levels of per capita income and different balance of payments problems. We had to address such issues as how nations should share the United Nations budget, how nations should share in the endowment of the International Monetary Fund and the World Bank, how nations should share in the UN Relief and Rehabilitation postwar program, and the like. It looked as if a tradition were developing among nations of how to impose a kind of informal international income tax on nations, for the purpose of sharing burdens that had to be shared.

Then the cold war ended all that progress, so we now lack traditions, precedents, or any kind of formula for how to share common burdens. The nations emitting carbon dioxide differ immensely in their reliance on fossil fuels. It is partly a question of which other fuels are available and whether the more carbon-intensive coal is more available than the less carbon-intensive natural gas. Reliance on fossil fuels also depends on the countries' climate and population density. People in Japan do not drive as many miles as Americans do because they have fewer places to drive.

Finding a way to agree on how all nations of the world, or a subset of those nations, might apportion reductions in carbon emissions is a real challenge. The costs of doing something serious about carbon emissions are variously estimated at 2 or 3 percent of gross national product, in perpetuity. That is to say, the annual lost production or the annual budgeted cost might be 2 or 3 percent of GNP.

If we were eliminating the federal budget deficit in this country, or in France, 2 percent of GNP would be an unmanageable burden. If I could wave a wand and phase in over the next ten years something that would reduce GNP per capita in the United States forever, by 2 percent below what it would otherwise have been, the difference, in a graph drawn on an $8^1/_2$-by-11-inch sheet of paper, would be about the thickness of a line drawn with a number-two pencil.

If I could bring about that 2 percent reduction by waving a wand, we would not notice it. But if I do it by taxing gasoline, everyone would notice it. For that reason, it is a larger problem politically than economically. Reducing GNP per capita by 2 percent over the next ten years would mean that the GNP per capita we might have reached by 2060 would not be realized until 2061 or 2062; the range of uncertainty, though, would swamp any such efforts.

Only one precedent for such a division of responsibility

among nations comes to mind: the Marshall Plan to rebuild Europe after World War II. Perhaps we can learn something from that success. The United States determined how the aid was to be handled in the period 1948–1949. For 1949–1950, there was about $4 billion to be divided among the participating countries of Western Europe, quite a lot of money back then. Depending on the exchange rate, that amount was at least 2 percent of GNP for every country concerned and may have been 5, 6, 7, or even 8 percent of GNP for some of the countries, both because in some cases the amount of aid was, in the end, larger and because their exchange rates vastly understated the domestic value of what they were getting.

In the late 1940s and early 1950s, those countries were dividing up something comparable to the order of magnitude of any program for drastically reducing carbon emissions over the next few decades.

How did they divide the aid? They followed a procedure for which I coined the phrase *multilateral reciprocal scrutiny*. Each country had to lay out, in a variety of tables and forms, exactly what it expected its own production to be; exactly what it would need from abroad; what it could get from other European countries; what it would have to pay for in dollars; what its plans were for private investment, for public investment, and for consumption; what it was doing about meat and gasoline rationing; and what it was doing about investing in increases in livestock or investing in transport and electric power facilities. Each participating country had to lay out all its facts and figures and then send senior officials to Paris to defend what they proposed to do and what share of aid they claimed in consequence.

This was an orderly, gentlemanly process of cross-examination that went on for a period of months, with the object of arriving at how they would divide $4 billion among about fifteen countries.

They did not quite come to a consensus. They were close enough to agreement, however, that they could select two disinterested people, who then went away for the weekend and came back with a proposed division that was immediately acceptable to all the countries.

Approximately the same thing happened with NATO, when the NATO buildup replaced European recovery as the object of U.S. aid to Europe. We changed the name from the European Recovery Program to the Mutual Defense Program but continued exactly the same kinds of activities. And those in Paris who had talked about dividing up the aid now talked about dividing up the military burdens: who would conscript how many soldiers, for how many months, and who would spend how much on military equipment, training, and supplies. It was essentially the same ballgame in the same ballpark with the same ballplayers and a new name. The same people went through the procedure of multilateral reciprocal scrutiny. And, again, they did not quite reach agreement on an allocation of military burdens or shares of aid. But they were close enough—this time three people went away and came back with an acceptable division.

Of course, the United States could have coerced those countries to find a way to come to agreement because we said no agreement, no aid. We may not have meant it, but at least we were there. In the present case, we have no such benefactor saying to all the countries that burn carbon: we will withhold something terribly important unless you come to agreement. We have nothing analogous to the United States behind the process of figuring out how to apportion the costs of improving the environment.

Beyond the Marshall Plan and NATO, I can think of no other occasion in history when national governments have had to negotiate shares in something that was on the order of a few percent of GNP. Negotiating how to divide assets is really no different from negotiating how to divide liabilities. We could say either that we are dividing up emission rights for carbon dioxide or that we are dividing up emission reductions. Either way, we have the same kind of process.

International Cooperation?

In 1996, 2,000 American economists, including a significant number of Nobel laureates, proposed that nations should get together and agree on fair quotas for carbon emissions and set up a system of emission trading rights to ensure an economical geographical distribution of the reductions.

I did not sign the statement, because I cannot imagine how the nations of the world could send representatives to a conference that would divide up the equivalent of a trillion dollars or more of trading rights. How could such a division be enforced? The U.S. Senate is unlikely to accept international enforcement of a procedure that would require the United States, if it burned more carbon than it was supposed to, to pay out tens of billions of dollars for the privilege. But also I do not believe that national representatives could agree on defining an ultimate emissions target. And unless they can set an ultimate target, then everybody will know that any agreement is temporary, subject to renegotiation. Moreover, if it were known that the allotments would be renegotiated in five or ten years, countries would be unlikely to sell their excess emission rights, because those rights would be clear evidence that the countries were dealt a better hand than they deserved at the beginning. I am thus skeptical of success.

Eventually, people will have to learn to talk not about emissions but about concentrations. Reducing emissions is a way to reach a climatically tolerable concentration of carbon dioxide, or all greenhouse gases, in the atmosphere. The Intergovernmental Panel on Climate Change, which comprises hundreds of scientists who collaborate regularly to study the subject and issue reports, has given no indication of what the climatically tolerable concentration in the atmosphere should be. They talk about concentrations from 450 to 750 parts per million. The concentration is now about 360 parts per million.

Until we have a good idea of what the ultimate concentration should be, it will be impossible to have any long-range planning. It is also undoubtedly the case that if we knew what our goal should be, we would increase emissions for the next few decades, before forcing them to taper off and then decline—maybe drastically—during the second half of the next century.

Climatically, it would make virtually no difference what the trajectory was by which we got to that ultimate concentration, while economically, it matters a great deal how we get there. I would advocate not delaying emissions reductions but rather reducing them less in the near term than in the far term. That approach does not mean doing nothing now, but it might mean spending a lot of money now developing the technologies that we will need in forty to fifty years to do what may become necessary. President Clinton's notion that $5 billion over five years would make any difference is preposterous.

Carbon Emissions and Foreign Aid

Eventually, we must deal with the Chinese. The Chinese now emit about half as much carbon as the United States, 12 to 13 percent of total global emissions. By the middle of the next century, they will probably be emitting a lot more than the United States is emitting now. In other words, their emissions are likely to triple or quadruple in the next fifty years, because they have vast plans to expand their electric power capability and they are very rich in coal, the most carbon-intensive fuel. And if we want to keep the Chinese from letting their emissions grow, and if nobody wants to consider invading China along the way, we must consider how to persuade them to find a way to reduce their emissions. If we want them to do anything in the short run—that is, within twenty-five years—we have to pay for it. If we want them to do something in the slightly longer run, we may persuade them.

What we want the Chinese to do will be difficult. Perhaps we will want to cover the extra cost of their building nuclear electric power plants instead of coal-fired power plants. But whatever we do, it will cost us money: unlike reducing emissions in the developed countries, which will probably come about more by regulation than by investment, to do anything about the Chinese or Indian or Indonesian or Nigerian emissions probably means budgeting funds. Such funds will look like foreign aid. But if we spend money to reduce Chinese emissions, for example, it will be foreign aid for any country that might benefit from reduced carbon emissions.

Organizing such aid will be tough. But we have more than a few years, maybe even a few decades, to get organized. If the countries of the European Union, the United States, Canada, Japan, and a few others can begin to get their own systems for reducing emissions in place over the next decade, we can then consider how to transfer resources to the developing world to help them reduce emissions, preferably in a way that does not leave room for blackmail on the part of developing countries.

I worry about one nagging problem that no one else appears to have considered: that is, the beneficiaries of anything we do to slow climate change will be the descendants of people now living in the poor countries. Those descendants will almost certainly be a lot better off than their grandparents. The question is, Why should we help future generations rather than those now living? The current generations are much poorer than their grandchildren and great-grandchildren will be.

There are two issues here. First, might we do more to protect the Indian population of 2050 and 2075, for example, from climate change by accelerating economic development in India now than by slowing down climatic change itself? There are many ways of reducing emissions that are cost effective enough to belong in a conservation program now. There may also be many ways of accelerating development so that if and when climate change comes about, people will be much less vulnerable to the damage that might occur.

Even for the grandchildren of those now in India, China, Indonesia, and other developing countries, putting a lot of resources into slowing climate change and nothing into their own accelerated development may be the wrong way to go. If I am wrong on that point, if we want to help the poor now and in the future, the more urgent need may be improving the quality of life of the people who are here now, rather than those fifty or seventy-five years from now. If today we had foreign aid to divide between Bangladesh and Singapore, who would give any to Singapore? But if many developing countries in fifty or seventy-five years will be close to Singapore's level of development now, then it seems backwards to avoid promoting economic development around the world today and focusing on slowing down climate change because of the good it will do for future generations.

The need for abatement of greenhouse gases should not be separated from the poor countries' need for rapid development. The trade-off between development and greenhouse emissions ought to be faced. But it probably will not happen soon.

Discussion

QUESTION: Mr. Schelling, from your argument, is it fair to say that we shouldn't be promoting development in the third world but subsidizing consumption instead?

PROFESSOR SCHELLING: If we want to improve the health of those in the third world, subsidizing clean water, sewage systems, a certain amount of transport, clinics and hospitals, vaccination programs, and efforts to control vector-borne diseases might make sense. Some of those activities will come under the heading of consumption, many of them under the heading of investment.

But the question also arises whether we would do more good in twenty years by sending in food now or by trying to increase food production. It probably varies from country to country, so we would want to do both.

QUESTION: Could you address the development and the climate problem simultaneously, recognizing that the biggest environmental problem and perhaps the largest social problem the world may face over the next century is the rapid urbanization of the developing world? In 1950, 285 million people lived in the cities throughout the world; in 1990, 1.4 billion. Some project that by 2025, we could have as many as 4 billion people in urban areas. If the international community were to focus on bringing electricity to the 2 billion people off the electric grid—those now rushing into cities because only there can they make a living—we could accomplish two things. Increasing electrification is likely to bring down the price of electricity substantially as the volume goes up, thereby supporting economic development, at the same time that people have less need to leave rural areas to find jobs. Development will occur in a much more benign form than would be brought about by an extension of conventional electrofits.

We may thus be able to reduce both problems and avoid some negative trade-offs.

PROFESSOR SCHELLING: We have to be careful about the term *urbanization*. I think the movement from the countryside to urban areas is inevitable. I think you are worried that everyone is moving to Beijing, Mexico City, or New Delhi. That is an agglomerative problem.

We need to develop urban centers where people will converge as they move from the countryside—places that are not simply the periphery of the megalopolises. I don't know how to accomplish that, but if it took some foreign aid to help countries learn how to develop their transport network, their electricity network, and various other facilities—including public health facilities—so that everybody wouldn't move into the large agglomerate areas, that would be very worthwhile.

As far as I can tell, the Chinese make some effort to keep people away from Shanghai, Beijing, and other large urban areas. How successful they are, I don't know. They probably employ techniques that would not be acceptable in India, Pakistan, Indonesia, Brazil, and elsewhere.

One of the advantages of trying to plan fifty years ahead with respect to infrastructure is that we have time to make sure that this centripetal force you're worried about can be offset. Urbanization can then pursue its inevitable course without necessarily resulting in more and more megalopolises.

QUESTION: You commented that recent models of world climate conditions predict less change in temperature than earlier models. The question still remains, though, whether there is a human-induced effect that would be higher than any natural effect. How would you respond?

PROFESSOR SCHELLING: I believe that none of these uncertainties will be resolved in the next few years; many may be resolved in the next decade or two. If we have 360 parts carbon gases per million now, by the time we get to 400 parts per million, if we still don't see a significant detectable change, we may know something we didn't know fifteen years ago.

I am not disappointed that no one is making progress on reducing emissions now because I think that within the next twenty years, we will have a better scientific understanding of the effects of carbon emissions on the global climate. We may also have a better way to communicate honestly to the public what they should be worrying about.

On the one hand, we cannot now rouse the American public to make any significant sacrifices for the sake of climate without being dishonestly alarmist. On the other hand, my rather benign estimate of what happens to material welfare as a result of climate change could be wrong. Some think it possible that ocean currents could be affected in some discontinuous fashion. The deep-water formation that occurs between Greenland and Iceland may be affected and may even disappear, in which case the Gulf Stream would have no place to go and would stop keeping Northern Europe warm. That is something to include in the total picture of whether we want to buy some "global climate insurance."

When I first heard about this subject around 1977, some predicted that global warming would cause the West Antarctic ice sheet—a body of ice within the ocean but grounded on the bottom—to glaciate or collapse into the ocean and raise the sea level by twenty feet in less than a century. It's not hard to estimate how much water-equivalent ice is there that would go into the ocean. The question was,

Would that happen and, if so, when would it happen?

Fortunately, satellite sensing of glacial movements was just then becoming available. Little beacons placed in the glaciers transmit data that satellites could quickly and reliably measure to show how rapidly various glaciers moved. Within a few years, it was decided that if the West Antarctic ice sheet were to disappear and raise ocean levels, it would probably take 500 years, rather than 100 years. A genuinely alarming possibility turned out, upon further investigation, not to be nearly as bad as originally forecast. Perhaps as we learn more about what determines ocean currents, we'll be able to dismiss the possibility that Scotland will freeze as a result of global warming.

COMMENT: I believe that we should move away from fossil fuels production and increase the share of renewable energy as quickly as possible. Our goal should be to make the world more energy efficient. If we do it early on, then the cost will be much less than if we do it later.

PROFESSOR SCHELLING: I disagree. I can cite at least four reasons why it makes more sense to go slow initially. First, it saves money to let existing capital investment go through its normal life and be retired. For my automobile, for example, that may be another six years, but for a new coal-fired electric power plant it would be another thirty or forty years. To shut it down in the interest of replacing it with nuclear power would be expensive. To shut it down anytime before its useful life is over would be wasteful. Converting some coal-fired plants to natural gas makes sense but maybe not for the plants that will be retired within ten or fifteen years.

Working through our capital uses capital resources wisely. As for increased energy efficiency in buildings, retrofitting may be harder than constructing new buildings in more energy-conserving ways.

Second, by postponing spending, we save the interest

on the expenditure. If we wanted to build a bridge across an estuary for fear that rising sea levels would require a higher bridge, for example, we might ask, "If it costs more to build a bridge that is ten feet higher than the bridge we planned, would it be cheaper to put the money in the bank and let it earn interest, and retrofit the bridge thirty or so years from now?" Saving interest matters.

Third, we will probably have a much better understanding of technology some decades from now. It's possible that the people currently apprehensive about nuclear power will be satisfied that we will have inherently safe nuclear power in the future.

Finally, we will be a lot richer in thirty or forty years. We are getting richer all the time. In a way, letting our children pay for adaptations to climate change may make sense if they are going to be better off than we are.

QUESTION: What if the developed nations agreed to a scheme to set the price of an emissions permit at one dollar, growing by a dollar a year for some specified term. Given the fact that, on balance, we will probably have to spend some money sometime on reduction of greenhouse gases, maybe we should get people to start talking to each other. Let's move away from establishing a fixed quantity of greenhouse gas emissions and agree to pay just a small price that will grow slightly over time and renegotiate the agreement in ten years. Is that a better way to look at it?

PROFESSOR SCHELLING: By "pay a price," do you mean have the U.S. government pay a penalty on emissions or have the consumer pay a price for fuel?

COMMENT: In effect, the government would agree to issue unlimited permits at a dollar apiece. Next year, the permit would cost two dollars.

PROFESSOR SCHELLING: Senator Byrd sponsored a resolution

in the Senate that was unanimously adopted: the president should agree to nothing at the 1997 Kyoto conference in which countries in the developing world were not full participants. The developing world, of course, will not participate in any plan that means reducing emissions back to 1990 levels by any time in the foreseeable future. Therefore, if you talk to Senator Byrd about using price as an indirect way of closing West Virginia coal fields, he will see through it.

I was John Anderson's energy adviser during his presidential campaign in 1980. He was already pushing the fifty-cent gasoline tax before I joined his team. My advice was that the tax was a splendid idea for coping with an energy crisis, but I was a little worried that it wouldn't get him elected—and it didn't. Jimmy Carter wanted a fifty-cent tax on motor fuel. He didn't get anywhere either.

Just a few years ago, the price of diesel fuel suddenly jumped eighteen cents. Truckers from all over the eastern United States converged on Washington, D.C., and blocked the bridges across the Potomac River in protest, not against anything Washington had done but against the price of diesel fuel.

It isn't easy to put anything over on Congress by claiming that we'll start with just a little tax and increase it over time. I think we have to come cleaner than that—and we can't until we have a very good idea of where we want to go.

My proposal would be to add a dime per gallon to the price of gasoline every year for fifteen years, so that as the tax slowly rises to a dollar and a half, people will anticipate the higher price of gasoline as they buy new cars. I see two problems with my proposal, though. First, you can't legislate a permanent tax; Congress can always turn around and change its mind. Second, no one can really predict what higher gas prices would do to carbon emissions by the year 2020. I've had a hard time finding any economists who could help me figure it out.

We experimented a bit with higher gasoline prices during the energy crisis of the middle 1970s. The crisis didn't last long enough for us to learn about the long-run elasticity of demand for gasoline, diesel fuel, or heating fuel. When European governments, the U.S. government, and the Japanese government seriously address greenhouse gas emissions, I think they should avoid what are now called "targets" and "timetables." They should be thinking about action programs and about what they want to tax, what they want to subsidize, what research they want to fund, and what they want to regulate. They must recognize that although they can then estimate what such activities will do to emissions, their estimates will be very poor.

I will consider governments serious when they talk about what actions they will take rather than what greenhouse emissions targets they hope to reach in future years.

About the Author

THOMAS C. SCHELLING is Distinguished University Professor at the University of Maryland and until 1990 was Lucius N. Littauer Professor of Political Economy at Harvard University. He is a member of the National Academy of Sciences and the Institute of Medicine and a fellow of the American Academy of Arts and Sciences. He was a member of the National Academy's Carbon Dioxide Assessment Committee and is currently on that academy's Commission on Geosciences, Environment, and Resources. He was chosen a distinguished fellow of the American Economic Association and elected president of that association for 1991. He received the Frank E. Seidman Distinguished Award in Political Economy and the National Academy of Sciences Award for Behavioral Research Relevant to the Prevention of Nuclear War. He is the author of eight books and 175 articles.

Board of Trustees

Wilson H. Taylor, *Chairman*
Chairman and CEO
CIGNA Corporation

Tully M. Friedman, *Treasurer*
Tully M. Friedman & Fleicher, LLC

Joseph A. Cannon
Chairman and CEO
Geneva Steel Company

Dick Cheney
Chairman and CEO
Halliburton Company

Harlan Crow
Managing Partner
Crow Family Holdings

Christopher C. DeMuth
President
American Enterprise Institute

Steve Forbes
President and CEO
Forbes Inc.

Christopher B. Galvin
CEO
Motorola, Inc.

Harvey Golub
Chairman and CEO
American Express Company

Robert F. Greenhill
Chairman
Greenhill & Co., LLC

Roger Hertog
President and COO
Sanford C. Bernstein and Company

M. Douglas Ivester
Chairman and CEO
The Coca-Cola Company

Martin M. Koffel
Chairman and CEO
URS Corporation

Bruce Kovner
Chairman
Caxton Corporation

Kenneth L. Lay
Chairman and CEO
Enron Corp.

Marilyn Ware Lewis
Chairman
American Water Works Co., Inc.

Alex J. Mandl
Chairman and CEO
Teligent, LLC

The American Enterprise Institute for Public Policy Research

Founded in 1943, AEI is a nonpartisan, nonprofit, research and educational organization based in Washington, D. C. The Institute sponsors research, conducts seminars and conferences, and publishes books and periodicals.

AEI's research is carried out under three major programs: Economic Policy Studies; Foreign Policy and Defense Studies; and Social and Political Studies. The resident scholars and fellows listed in these pages are part of a network that also includes ninety adjunct scholars at leading universities throughout the United States and in several foreign countries.

The views expressed in AEI publications are those of the authors and do not necessarily reflect the views of the staff, advisory panels, officers, or trustees.

Craig O. McCaw
Chairman and CEO
Eagle River, Inc.

Paul H. O'Neill
Chairman and CEO
Aluminum Company of America

John E. Pepper
Chairman and CEO
The Procter & Gamble Company

George R. Roberts
Kohlberg Kravis Roberts & Co.

John W. Rowe
President and CEO
New England Electric System

Edward B. Rust, Jr.
President and CEO
State Farm Insurance Companies

John W. Snow
Chairman, President, and CEO
CSX Corporation

William S. Stavropoulos
Chairman and CEO
The Dow Chemical Company

Officers

Christopher C. DeMuth
President

David B. Gerson
Executive Vice President

John R. Bolton
Senior Vice President

Council of Academic Advisers

James Q. Wilson, *Chairman*
James A. Collins Professor of Management Emeritus
University of California at Los Angeles

Gertrude Himmelfarb
Distinguished Professor of History Emeritus
City University of New York

Samuel P. Huntington
Eaton Professor of the Science of Government
Harvard University

D. Gale Johnson
Eliakim Hastings Moore Distinguished Service Professor Economics Emeritus
University of Chicago

William M. Landes
Clifton R. Musser Professor of Economics
University of Chicago Law School

Sam Peltzman
Sears Roebuck Professor of Economics and Financial Services
University of Chicago Graduate School of Business

Nelson W. Polsby
Professor of Political Science
University of California at Berkeley

·ge L. Priest
M. Olin Professor of Law and
omics
aw School

nas Sowell
r Fellow
er Institution
ord University

·ay L. Weidenbaum
nckrodt Distinguished
iversity Professor
ington University

ard J. Zeckhauser
Ramsey Professor of Political
nomy
edy School of Government
ard University

earch Staff

 Aron
ent Scholar

de E. Barfield
ent Scholar; Director, Science
Technology Policy Studies

·hia A. Beltz
·rch Fellow

er Berns
ent Scholar

·las J. Besharov
nt Scholar

·t H. Bork
1. Olin Scholar in Legal Studies

·n Bowman
nt Fellow

·eth Brown
·g Fellow

·E. Calfee
·nt Scholar

· V. Cheney
 Fellow

·h D'Souza
1. Olin Research Fellow

·las N. Eberstadt
·g Scholar

 Falcoff
·nt Scholar

·d R. Ford
·uished Fellow

Murray F. Foss
Visiting Scholar

Michael Fumento
Resident Fellow

Diana Furchtgott-Roth
Assistant to the President and
  Resident Fellow

Suzanne Garment
Resident Scholar

Jeffrey Gedmin
Research Fellow

James K. Glassman
DeWitt Wallace–Reader's Digest
  Fellow

Robert A. Goldwin
Resident Scholar

Mark Groombridge
Abramson Fellow; Associate Director,
  Asian Studies

Robert W. Hahn
Resident Scholar

Kevin Hassett
Resident Scholar

Robert B. Helms
Resident Scholar; Director, Health
  Policy Studies

R. Glenn Hubbard
Visiting Scholar

James D. Johnston
Resident Fellow

Jeane J. Kirkpatrick
Senior Fellow; Director, Foreign
  and Defense Policy Studies

Marvin H. Kosters
Resident Scholar; Director,
  Economic Policy Studies

Irving Kristol
John M. Olin Distinguished Fellow

Dana Lane
Director of Publications

Michael A. Ledeen
Freedom Scholar

James Lilley
Resident Fellow

Clarisa Long
Abramson Fellow

Lawrence Lindsey
Arthur F. Burns Scholar in Economics

John H. Makin
Resident Scholar; Director, Fiscal
  Policy Studies

Allan H. Meltzer
Visiting Scholar

Joshua Muravchik
Resident Scholar

Charles Murray
Bradley Fellow

Michael Novak
George F. Jewett Scholar in Religion,
  Philosophy, and Public Policy;
  Director, Social and Political Studies

Norman J. Ornstein
Resident Scholar

Richard N. Perle
Resident Fellow

William Schneider
Resident Scholar

William Shew
Visiting Scholar

J. Gregory Sidak
F. K. Weyerhaeuser Fellow

Christina Hoff Sommers
W. H. Brady, Jr., Fellow

Herbert Stein
Senior Fellow

Irwin M. Stelzer
Resident Scholar; Director,
  Regulatory Policy Studies

Daniel Troy
Associate Scholar

Arthur Waldron
Director, Asian Studies

W. Allen Wallis
Resident Scholar

Ben J. Wattenberg
Senior Fellow

Carolyn L. Weaver
Resident Scholar; Director, Social
  Security and Pension Studies

Karl Zinsmeister
J. B. Fuqua Fellow; Editor, *The
  American Enterprise*

AEI STUDIES ON GLOBAL ENVIRONMENTAL POLICY
Irwin M. Stelzer, Series Editor

Costs and Benefits of Greenhouse Gas Reduction
Thomas C. Schelling

The Economics and Politics of Climate Change
Robert W. Hahn

Making Environmental Policy: Two Views
Irwin M. Stelzer and Paul R. Portney

www.ingramcontent.com/pod-product-compliance
Lightning Source LLC
Jackson TN
JSHW011944131224
75386JS00041B/1557